# Journey to the Infinite: Exploring the Mysteries of the Universe

Introduction:

Welcome to "Journey to the Infinite: Exploring the Mysteries of the Universe"! This book is a journey through the vast and fascinating world of astronomy, taking you on a tour of the mysteries and wonders of the universe.

From the tiniest particles to the largest structures, the universe is a vast and complex place. This book will introduce you to the fundamental principles of astronomy, including the laws of physics that govern the cosmos, the history of our universe, and the latest scientific discoveries.

You will discover how astronomers study distant stars and galaxies, and learn about the incredible instruments they use to observe the universe. We will explore the origins of the universe, from the Big Bang to the formation of stars and galaxies, and examine some of the biggest questions in astronomy, such as the search for life beyond our planet.

The book will also delve into some of the most fascinating phenomena in the universe, from black holes and dark matter to supernovae and the birth of new stars. You will learn about the different types of galaxies and their evolution over time, and discover how the universe is expanding and changing.

Whether you are a student of astronomy, a science enthusiast, or simply curious about the mysteries of the universe, "Journey to the Infinite" will take you on a captivating journey through the cosmos. With clear explanations, stunning visuals, and thought-provoking insights, this book is the perfect introduction to the wonders of the universe.

# Index

1. Discover the fundamental principles of astronomy.

2. Explore the history of the universe, from the Big Bang to the present day.

3. Learn about the incredible instruments used by astronomers to study the cosmos.

4. Investigate the search for life beyond our planet.

5. Delve into the fascinating phenomena of black holes, dark matter, and supernovae.

6. Examine the different types of galaxies and their evolution over time.

7. Gain insights into the expanding and changing universe.

8. Experience a captivating journey through the wonders of the cosmos.

## Chapter-1

# Discover the fundamental principles of astronomy

Astronomy is the scientific study of celestial objects, such as stars, planets, galaxies, and other celestial bodies, and the physical and chemical processes that occur within them. The following is a summary of the fundamental principles of astronomy.

1. **The universe is vast**: The universe is an inconceivably large expanse of space and time, with billions of galaxies, each containing billions of stars. The observable universe is estimated to be about 93 billion light-years in diameter, although the entire universe may be even larger.

2. **Everything is in motion**: All celestial objects, including planets, stars, and galaxies, are in motion. The planets orbit around their respective stars, and stars orbit around the centers of their galaxies. This motion is described by the laws of physics, including

Kepler's laws of planetary motion and Newton's laws of motion.

3. **Light is crucial**: Light is the primary means by which astronomers gather information about the universe. The electromagnetic spectrum, which includes visible light, radio waves, and X-rays, is used to study celestial objects. Telescopes, both ground-based and space-based, are used to capture and analyze light from distant objects.

4. **Gravity governs the universe**: Gravity is the force that governs the motion of celestial objects. It is the force that holds planets in orbit around their stars, keeps stars in their galaxies, and causes galaxies to cluster together. Einstein's theory of general relativity provides a more complete understanding of gravity and its effects on the universe.

5. **The big bang**: The current scientific consensus is that the universe began with the Big Bang, a massive explosion that occurred about 13.8 billion years ago. The Big Bang

theory explains the observed properties of the universe, including its large-scale structure, the cosmic microwave background radiation, and the abundance of light elements.

6. **Evolution of stars and galaxies**: Stars and galaxies evolve over time. Stars are born from clouds of gas and dust, and they undergo various stages of fusion and nuclear reactions before eventually dying and releasing their material back into the universe. Galaxies form from the collision and merging of smaller structures, and they evolve through interactions with other galaxies.

7. **Dark matter and dark energy**: The universe contains vast amounts of matter and energy that cannot be directly observed, known as dark matter and dark energy. These substances are inferred from their gravitational effects on visible matter, and they make up the majority of the universe's mass-energy budget.

8. **Search for life beyond Earth**: Astronomy also explores the possibility of life

beyond Earth. The search for extraterrestrial intelligence (SETI) involves listening for signals from other civilizations in the universe, while the search for exoplanets involves identifying planets around other stars that may be capable of supporting life.

**In conclusion**, astronomy is a fascinating field that provides a deep understanding of the universe and its workings. Its fundamental principles are based on the laws of physics, and it continues to expand our knowledge of the cosmos.

# Chapter-2

Explore the history of the universe, from the Big Bang to the present day

The history of the universe is the story of the evolution of the cosmos from its origin in the Big Bang to the present day. The following is a summary of the major events in the history of the universe.

1. **The Big Bang**: The universe began with the Big Bang, a massive explosion that occurred about 13.8 billion years ago. At the moment of the Big Bang, the universe was infinitely hot and dense, and all matter and energy were compressed into a single point of infinite density known as a singularity. The universe rapidly expanded and cooled, and matter began to condense into galaxies and stars.

2. **Formation of galaxies and stars**: Over time, the matter in the universe began to clump together due to gravity, forming galaxies and stars. The first galaxies formed about 200 million years after the Big Bang, and the first stars formed shortly thereafter. These early stars were much larger and hotter than stars today, and they exploded in supernovae that seeded the universe with heavy elements.

3. **Formation of planets**: As stars formed, they were surrounded by disks of gas and

dust that eventually coalesced into planets. Our own solar system formed about 4.6 billion years ago, when a cloud of gas and dust collapsed under its own gravity. The inner planets, including Earth, are made of rock and metal, while the outer planets are composed of gas and ice.

4. **Formation of life**: The first life forms on Earth are believed to have appeared about 3.5 billion years ago. These early organisms were single-celled and lived in the oceans. Over time, life on Earth evolved into more complex forms, including multicellular organisms and eventually humans.

5. **Expansion of the universe**: The universe continues to expand to this day. This expansion is driven by dark energy, a mysterious force that causes space to stretch apart. The rate of expansion is accelerating, meaning that the universe is getting larger at an ever-increasing rate.

6. **Fate of the universe**: The ultimate fate of the universe depends on the amount of

matter and dark energy it contains. If there is enough matter, gravity will eventually slow the expansion of the universe and cause it to collapse back in on itself in a "Big Crunch." If there is not enough matter, the universe will continue to expand forever, eventually becoming a cold, dark, and empty place.

**In conclusion**, the history of the universe is a fascinating tale of the evolution of the cosmos from its explosive beginning to its uncertain future. From the formation of galaxies and stars to the emergence of life on Earth, the universe has undergone many transformations over its long and storied history. As we continue to study the universe, we are constantly uncovering new insights into its past, present, and future.

# Chapter-3

Learn about the incredible instruments used by astronomers to study the cosmos

Astronomers use a wide range of sophisticated instruments to study the cosmos, including telescopes, cameras, and spectrometers. These instruments allow astronomers to collect data from distant stars and galaxies, and to analyze the properties of these celestial objects.

1. **Telescopes**: Telescopes are the most important tool in an astronomer's toolkit. They are used to collect and focus light from distant objects, allowing astronomers to see faraway stars and galaxies. Telescopes can be either ground-based or space-based, with each type offering its own unique advantages. Some of the most powerful telescopes in use today include the Hubble Space Telescope, the Chandra X-ray Observatory, and the Atacama Large Millimeter/submillimeter Array (ALMA).

2. **Cameras**: Cameras are used in conjunction with telescopes to capture images of the night sky. These images can be

used to study the properties of stars and galaxies, and to track their movements over time. Modern cameras used in astronomy are incredibly sensitive, allowing astronomers to detect very faint sources of light.

3. **Spectrometers**: Spectrometers are used to study the properties of light emitted by celestial objects. They work by splitting light into its component wavelengths, which can then be analyzed to determine the object's temperature, chemical composition, and other properties. Spectrometers are used in a wide range of astronomical observations, from studying the atmospheres of exoplanets to analyzing the composition of distant galaxies.

4. **Radio telescopes**: Radio telescopes are used to study radio waves emitted by celestial objects. These telescopes are often used in conjunction with other instruments, such as cameras and spectrometers, to create detailed images of the night sky. Some of the most famous radio telescopes include the

Very Large Array (VLA) in New Mexico, USA, and the LOFAR telescope in the Netherlands.

5. **Interferometers**: Interferometers are used to combine data from multiple telescopes to create detailed images of the night sky. They work by combining the light collected by multiple telescopes to create a single, high-resolution image. Interferometers are used in a wide range of astronomical observations, from studying the structure of nearby galaxies to mapping the cosmic microwave background radiation left over from the Big Bang.

6. **Spacecraft**: Spacecraft are used to study the cosmos up close. They are equipped with a wide range of scientific instruments, including cameras, spectrometers, and particle detectors. Some of the most famous space missions in astronomy include the Voyager missions, which explored the outer solar system, and the Kepler mission, which discovered thousands of exoplanets.

**In conclusion**, the instruments used by astronomers to study the cosmos are incredibly sophisticated and powerful. From telescopes that can see billions of light-years into the past to spacecraft that explore the outer reaches of our solar system, these instruments allow astronomers to explore the mysteries of the universe in unprecedented detail. As technology continues to advance, we can expect even more incredible instruments to be developed, enabling us to learn even more about the cosmos and our place in it.

# Chapter-4

Investigate the search for life beyond our planet

The search for life beyond our planet is one of the most exciting and important fields of scientific research today. Scientists are using a variety of methods and instruments to search for signs of life on other planets, including the study of exoplanets, the search for biosignatures, and the exploration of our own solar system.

1. **Exoplanets**: Exoplanets are planets that orbit stars other than our sun. Scientists have discovered thousands of exoplanets in recent years, and many of them are believed to be within the habitable zone of their stars, where conditions may be suitable for life. Astronomers study exoplanets using a variety of methods, including transit photometry, which involves measuring changes in a star's brightness as a planet passes in front of it, and radial velocity, which involves detecting the gravitational wobble of a star caused by an orbiting planet.

2. **Biosignatures**: Biosignatures are signs of life that can be detected remotely, such as the presence of certain chemicals in a planet's atmosphere. Scientists are studying the potential biosignatures of exoplanets using a variety of techniques, including spectroscopy, which involves analyzing the light absorbed or

emitted by a planet to determine its chemical composition.

3. **Our Solar System**: While most of the search for life beyond Earth focuses on exoplanets, scientists are also exploring our own solar system for signs of life. Mars, for example, is a prime target for exploration, as it has a history of liquid water on its surface and may still have subsurface water today. Other potential targets for the search for life in our solar system include the icy moons of Jupiter and Saturn, such as Europa and Enceladus, which may have subsurface oceans of liquid water.

4. **SETI**: The Search for Extraterrestrial Intelligence (SETI) is a program that uses radio telescopes to search for signals from intelligent civilizations beyond our planet. While SETI has yet to detect any conclusive evidence of extraterrestrial intelligence, the program continues to search, using increasingly sophisticated instruments and techniques.

5. **Astrobiology**: Astrobiology is a multidisciplinary field that studies the origin, evolution, and distribution of life in the universe. It brings together scientists from a variety of fields, including astronomy, biology, chemistry, and geology, to explore the fundamental questions of life beyond Earth.

**In conclusion**, the search for life beyond our planet is a fascinating and rapidly developing field of scientific research. While we have yet to find conclusive evidence of life beyond Earth, the discovery of exoplanets within the habitable zone of their stars and the identification of potential biosignatures are exciting developments that bring us one step closer to answering the age-old question of whether we are alone in the universe. As our technology continues to advance and our understanding of the cosmos grows, we can expect even more exciting discoveries in the search for life beyond our planet.

# Chapter-5

# Delve into the fascinating phenomena of black holes, dark matter, and supernovae

Black holes, dark matter, and supernovae are all fascinating phenomena that have captured the imagination of scientists and the public alike. These phenomena are some of the most mysterious and enigmatic in the universe, and they continue to puzzle scientists to this day.

Black holes are perhaps the most mysterious of all astrophysical objects. They are formed when a massive star runs out of fuel and collapses in on itself, creating a region of space where the gravitational pull is so strong that nothing, not even light, can escape. The boundary around a black hole from which nothing can escape is called the event horizon.

Black holes are thought to play a crucial role in the evolution of galaxies, as their powerful gravitational forces can influence the movement of nearby stars and even entire galaxies. Scientists continue to study black holes to learn more about their properties and how they interact with the universe around them.

Dark matter is another mysterious substance that has puzzled scientists for decades. It is thought to make up a significant portion of the universe's mass, but it cannot be directly observed. Instead, scientists can infer its presence from the way that it affects the motion of galaxies and other astronomical objects.

One of the most exciting things about dark matter is that it has the potential to fundamentally change our understanding of the universe. If scientists can determine what dark matter is made of and how it interacts with other particles, it could revolutionize our understanding of the laws of physics.

Supernovae are among the most spectacular events in the universe. They occur when a massive star runs out of fuel and undergoes a catastrophic explosion, releasing an enormous amount of energy in the process. Supernovae can outshine entire galaxies and are responsible for producing many of the heavy elements that make up our world.

Scientists study supernovae to learn more about the processes that occur inside stars and to understand how these explosions can influence the formation and evolution of galaxies. Supernovae are also of interest because they can serve as "standard candles" that allow scientists to measure the distance to remote objects in the universe.

**In conclusion**, black holes, dark matter, and supernovae are all fascinating phenomena that continue to puzzle scientists and capture the public's imagination. Each of these phenomena offers unique insights into the workings of the universe, and scientists will undoubtedly continue to study them for years to come in the hopes of unlocking their secrets.

## Chapter-6

# Examine the different types of galaxies and their evolution over time

Galaxies are vast systems of stars, gas, and dust that are held together by gravity. There are several different types of galaxies, each with its own distinct characteristics and evolution over time.

The most common type of galaxy is the spiral galaxy. Spiral galaxies are characterized by a flat, rotating disk of stars and gas, with a central bulge and spiral arms that extend outward. The Milky Way is an example of a spiral galaxy. Spiral galaxies typically have a large amount of interstellar gas and dust, which can fuel star formation.

Another type of galaxy is the elliptical galaxy. Elliptical galaxies are shaped like elongated spheres and are composed of old stars, with little interstellar gas and dust. Elliptical galaxies are thought to be formed through the merging of several smaller galaxies, which disrupts the spiral structure and causes the stars to redistribute into a more spheroidal shape.

Irregular galaxies are another type of galaxy that do not have a well-defined shape. They are typically smaller than spiral and elliptical galaxies and have a higher rate of star formation. Irregular galaxies are often found in the vicinity of larger galaxies and are thought to be formed through gravitational interactions with other galaxies.

Over time, galaxies can undergo significant changes in their structure and composition. One of the most significant factors influencing galaxy evolution is the rate of star formation. Galaxies that are actively forming stars are said to be in a "starburst" phase and can undergo significant changes in structure and composition as a result.

Galaxies can also evolve through interactions with other galaxies. When two galaxies merge, the resulting galaxy can have a different structure and composition than the original galaxies. Galaxies can also experience tidal interactions with other galaxies, which can cause the gas and dust within the galaxy to be redistributed and alter the rate of star formation.

The evolution of galaxies over time is also influenced by the presence of dark matter. Dark matter is an invisible substance that is thought to make up a significant portion of the mass of the universe. The gravitational pull of dark matter can influence the structure and movement of galaxies, causing them to evolve in different ways.

**In conclusion**, galaxies are diverse and complex systems that come in many different shapes and sizes. Each type of galaxy has its own unique characteristics and evolution over time. Spiral galaxies are characterized by their flat, rotating disks of stars and gas, while elliptical galaxies are shaped like elongated spheres and contain mostly old stars. Irregular galaxies have no well-defined shape and are typically smaller than other types of galaxies. The evolution of galaxies is influenced by a variety of factors, including the rate of star formation, interactions with other galaxies, and the presence of dark matter. Scientists continue to study galaxies to better understand their formation and evolution over time.

# Chapter-7

Gain insights into the expanding and changing universe

The universe is constantly expanding and changing. Since the discovery of cosmic expansion in the 1920s, scientists have been working to understand the nature of this expansion and its implications for the evolution of the universe over time.

The expanding universe is driven by a mysterious force known as dark energy. This force is thought to be responsible for the acceleration of the universe's expansion, which was first discovered in the late 1990s through observations of distant supernovae.

Dark energy is thought to make up around 70% of the universe's energy density, although its nature is still largely unknown.

The expansion of the universe has important implications for the evolution of galaxies and other astronomical objects. As the universe expands, galaxies and other objects are moving away from each other at an increasing rate. This means that over time, galaxies will become more and more isolated from each other, making it increasingly difficult for stars to form and for galaxies to interact with each other.

In addition to the expansion of the universe, scientists have also discovered that the universe has gone through several periods of rapid change over its history. One of the most significant of these periods is known as inflation. Inflation occurred in the first fractions of a second after the Big Bang, when the universe expanded exponentially, stretching out to many times its original size in a fraction of a second.

The early universe was also characterized by a hot, dense plasma of particles known as the quark-gluon plasma. This plasma was formed in the moments after the Big Bang and lasted for only a few microseconds before cooling and expanding to form the matter that we see in the universe today.

As the universe has expanded and cooled over time, it has gone through several phase transitions. One of the most important of these transitions is known as the electroweak transition, which occurred around $10^{-12}$ seconds after the Big Bang. During this transition, the electromagnetic and weak forces were unified into a single force, which then later separated into the two distinct forces that we observe today.

The changing nature of the universe is also reflected in the evolution of galaxies and other astronomical objects. Galaxies have undergone significant changes over time, from the early formation of the first galaxies in the universe to the evolution of modern galaxies through processes such as mergers and star formation.

**In conclusion**, the expanding and changing nature of the universe is one of the most fascinating and important areas of research in astrophysics. The universe's expansion is driven by the mysterious force of dark energy, which has important implications for the evolution of galaxies and other astronomical objects. The universe has also gone through several periods of rapid change, from the inflationary period in the moments after the Big Bang to the electroweak transition and the formation of galaxies and other astronomical objects. As scientists continue to study the universe, they will undoubtedly uncover many more secrets about its evolution over time.

# Chapter-8

Experience a captivating journey through the wonders of the cosmos

The cosmos, or the universe as a whole, is a vast and mysterious place filled with wonders that have captivated humans for millennia.

From the glittering stars in the night sky to the massive black holes at the centers of galaxies, the cosmos is an endlessly fascinating place to explore.

One of the most exciting areas of research in astrophysics is the search for life beyond Earth. Scientists have identified thousands of exoplanets, or planets that orbit stars other than the Sun, and are working to understand the conditions that are necessary for life to exist. Some of the most promising places to search for life include the icy moons of Jupiter and Saturn, which may have subsurface oceans that could support life.

Another area of fascination in the cosmos is the study of black holes. Black holes are objects with such strong gravitational fields that nothing, not even light, can escape once it crosses the event horizon. Scientists are working to understand the nature of black holes, including how they form and evolve over time. Some of the most exciting discoveries in recent years have been made through the observation of gravitational waves, which are ripples in the fabric of space-time that are created by the collision of two black holes.

The cosmos is also home to many stunningly beautiful objects, such as nebulae and galaxies. Nebulae are clouds of gas and dust that are illuminated by nearby stars, creating breathtakingly colorful displays in the night sky. Galaxies are enormous collections of stars, gas, and dust that come in a variety of shapes and sizes. Scientists are working to understand the evolution of galaxies over time, including the formation of the first galaxies in the early universe and the merging of galaxies that can result in the creation of new stars and planets.

The study of the cosmos is not limited to visible light. Scientists also study the universe through the observation of other forms of electromagnetic radiation, including radio waves, X-rays, and gamma rays. These different forms of radiation can provide unique insights into the nature of the universe, including the behavior of high-energy particles and the properties of distant galaxies.

Finally, the cosmos is also home to some of the most mysterious and enigmatic phenomena in the universe. These include dark matter, which is thought to make up around 27% of the universe's energy density but has never been directly observed, and dark energy, which is thought to be responsible for the acceleration of the universe's expansion. Scientists are working to understand the nature of these mysterious substances, which hold the key to unlocking some of the deepest secrets of the cosmos.

**In conclusion**, the cosmos is a fascinating and endlessly intriguing place to explore. From the search for life beyond Earth to the study of black holes and the observation of stunningly beautiful nebulae and galaxies, the cosmos offers a wealth of opportunities for scientific exploration and discovery. As scientists continue to study the universe, they will undoubtedly uncover many more secrets about the nature of the cosmos and our place within it.

www.ingramcontent.com/pod-product-compliance
Lightning Source LLC
Chambersburg PA
CBHW041949240526
45473CB00036B/2792